AF305410

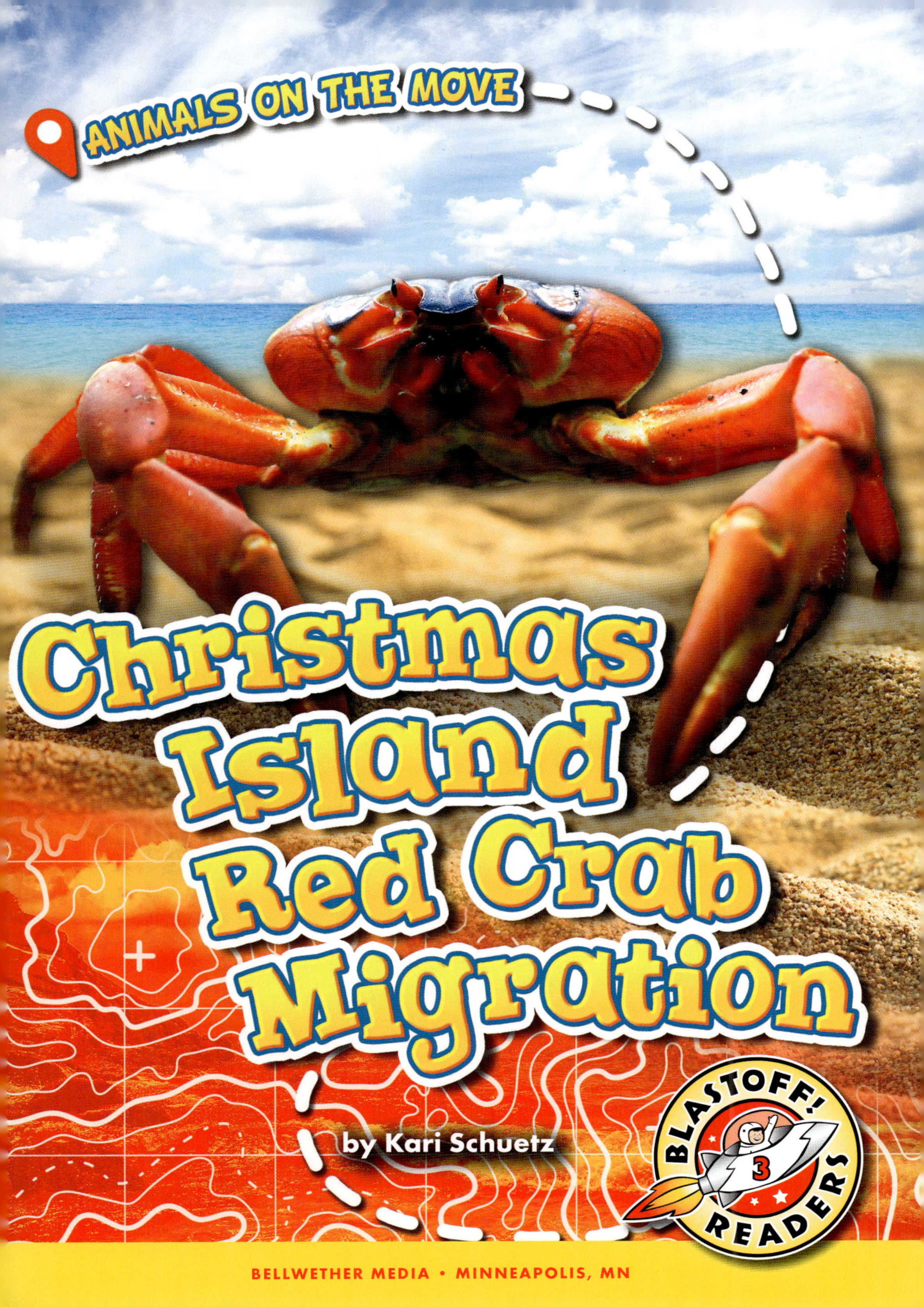

ANIMALS ON THE MOVE
Christmas Island Red Crab Migration
by Kari Schuetz
BLASTOFF! READERS
3
BELLWETHER MEDIA • MINNEAPOLIS, MN

Note to Librarians, Teachers, and Parents:

Blastoff! Readers are carefully developed by literacy experts and combine standards-based content with developmentally appropriate text.

Level 1 provides the most support through repetition of high-frequency words, light text, predictable sentence patterns, and strong visual support.

Level 2 offers early readers a bit more challenge through varied simple sentences, increased text load, and less repetition of high-frequency words.

Level 3 advances early-fluent readers toward fluency through increased text and concept load, less reliance on visuals, longer sentences, and more literary language.

Level 4 builds reading stamina by providing more text per page, increased use of punctuation, greater variation in sentence patterns, and increasingly challenging vocabulary.

Level 5 encourages children to move from "learning to read" to "reading to learn" by providing even more text, varied writing styles, and less familiar topics.

Whichever book is right for your reader, Blastoff! Readers are the perfect books to build confidence and encourage a love of reading that will last a lifetime!

This edition first published in 2019 by Bellwether Media, Inc.

No part of this publication may be reproduced in whole or in part without written permission of the publisher. For information regarding permission, write to Bellwether Media, Inc., Attention: Permissions Department, 6012 Blue Circle Drive, Minnetonka, MN 55343.

Library of Congress Cataloging-in-Publication Data

Names: Schuetz, Kari, author.
Title: Christmas Island Red Crab Migration / by Kari Schuetz.
Description: Minneapolis, MN : Bellwether Media, Inc., [2018] | Series: Blastoff! Readers. Animals on the Move | Audience: Ages 5-8. | Audience: K to grade 3. | Includes bibliographical references and index.
Identifiers: LCCN 2017061808 (print) | LCCN 2018000997 (ebook) | ISBN 9781626178151 (hardcover : alk. paper)| ISBN 9781681035567 (ebook)
Subjects: LCSH: Chaceon--Christmas Island (Indian Ocean)--Juvenile literature. | Gecarcinidae--Christmas Island (Indian Ocean)--Juvenile literature. | Crabs--Migration--Juvenile literature. | Animal migration--Juvenile literature.
Classification: LCC QL444.M33 (ebook) | LCC QL444.M33 S3675 2019 (print) | DDC 595.3/86--dc23

LC record available at https://lccn.loc.gov/2017061808

Editor: Paige V. Polinsky   Designer: Jeffrey Kollock

Printed in the United States of America, North Mankato, MN

Table of Contents

Christmas Island Red Crabs

Christmas Island red crabs are famous beach visitors. Every year, they swarm a small Australian island's shores.

Christmas Island Red Crab Profile

Millions of these bright **crustaceans** scurry toward the Indian Ocean to **breed**. Their **migration** stops traffic!

The small red crabs have hard shells. These **shields** keep them safe on their journey.

Fast, **nimble** legs carry them far.
Strong claws help the crabs dig
through soil and sand.

A Major March

Christmas Island red crabs live in the **rain forest**. The **wet season** draws them out of their **burrows**.

During this time, **mature** crabs can travel to the beach. Rainy weather lets them survive out in the open.

Male crabs begin
marching to the ocean.
Soon, females
join their parades.
The migration causes
them to crowd together.

This is a big change
for the crabs. During
the **dry season**,
they live alone.

Christmas Island
Red Crab Departure

mode of travel: walking

arriving
1 to 2 weeks after
leaving: coast

leaving
October/November:
rain forests

Crustacean Crossing

The crabs follow straight paths to the coast. These lead them across roads and down cliffs.

Christmas Island Red Crab Migration

It is a hard trip. Some crabs get hit by cars or fall from high places. Hungry ants attack others.

The crabs must stay wet
to survive. They seek shade
on sunny days.

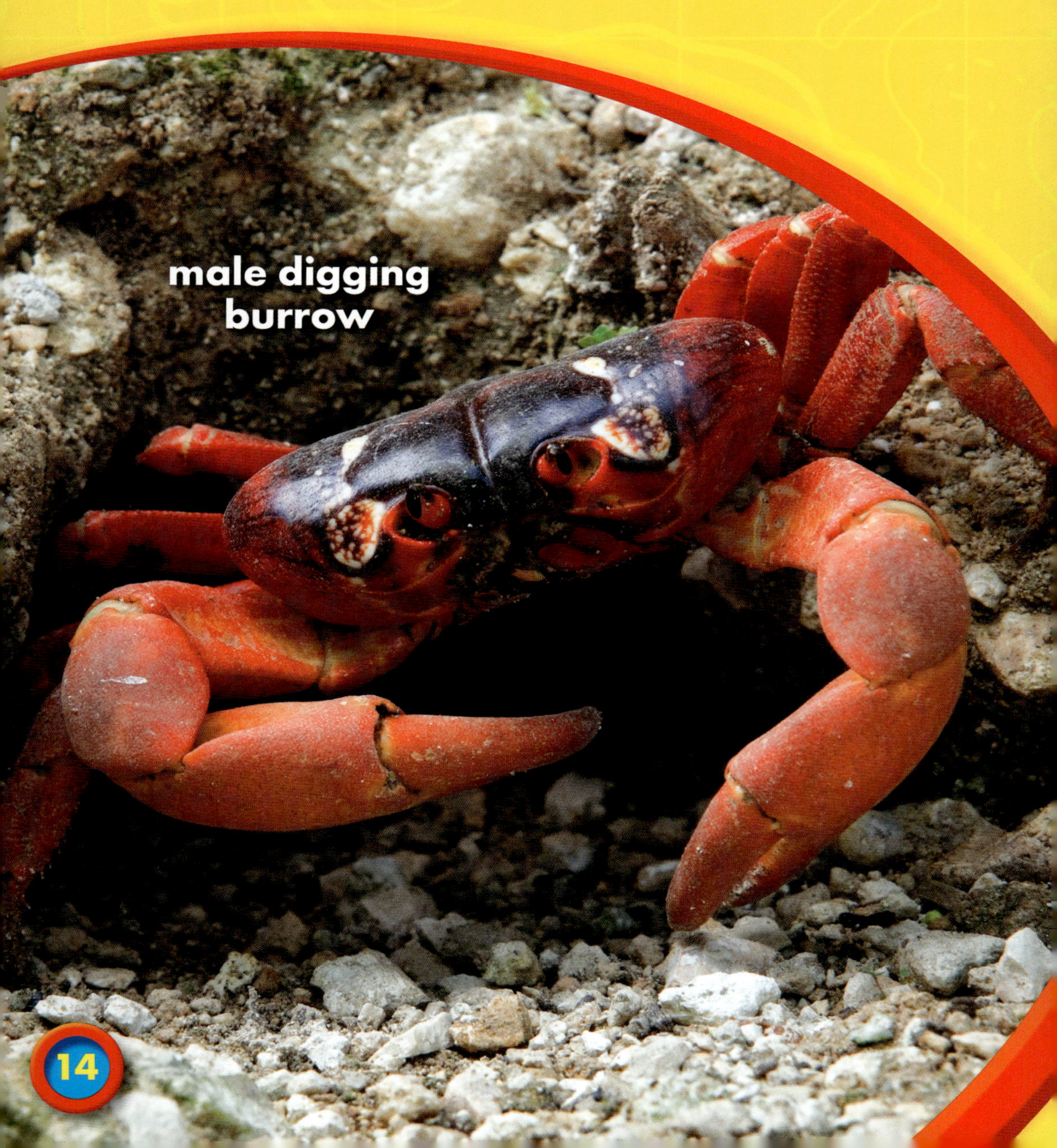

14

Once they reach the beach, they dip in the water right away. Then males dig burrows in the sand.

Males breed with females and then return to the rain forest. Females stay to lay eggs inside the burrows.

After a couple weeks, they move their eggs to the water. Then the females leave, too.

Babies on the Move

larvae

In the ocean, crab **larvae** quickly **hatch** from the eggs. They grow for about a month before heading to land.

Many baby crabs are eaten by **predators** before they reach shore. But lucky babies arrive safely.

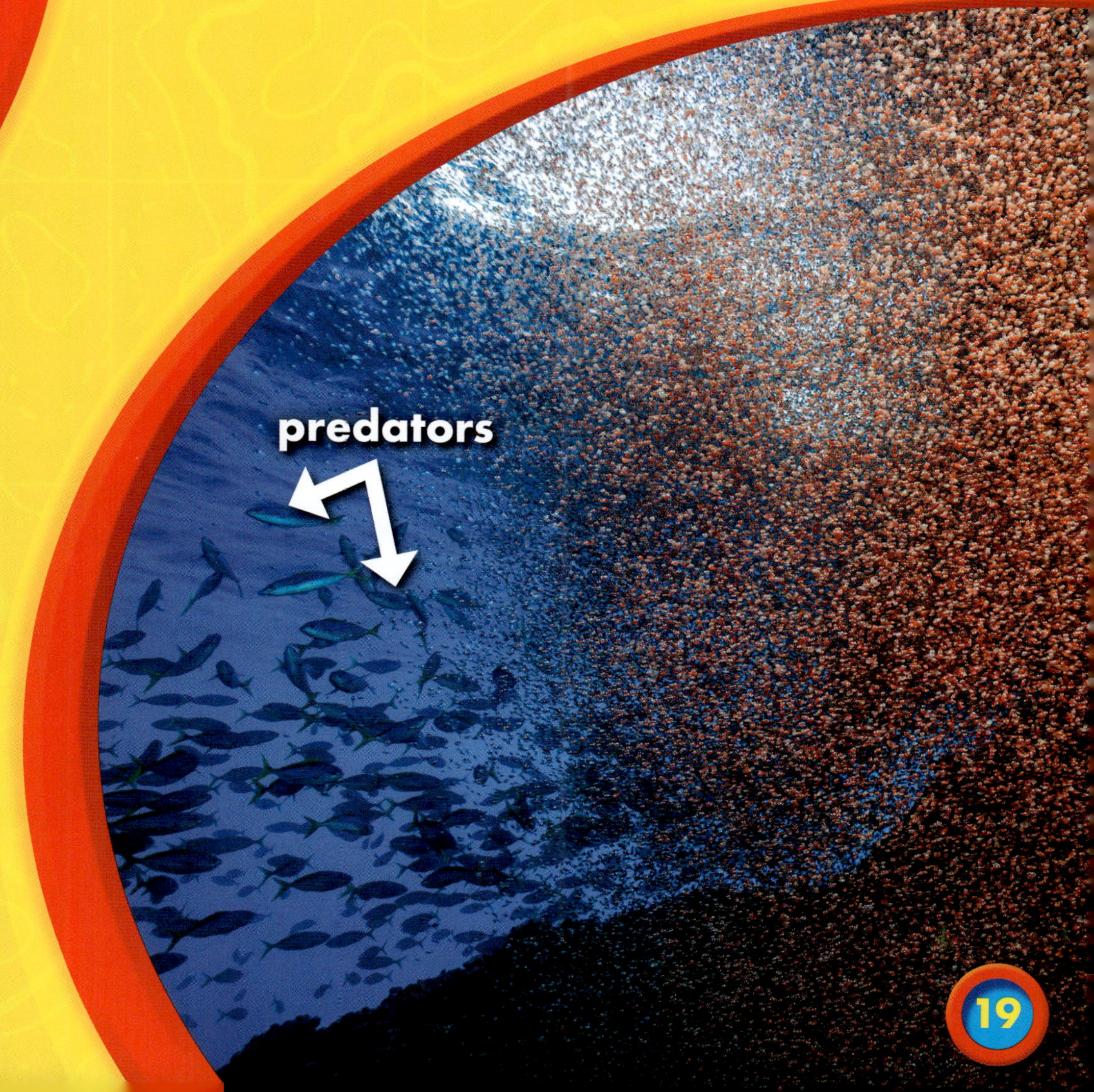

The baby crabs crawl their way to the rain forest. These tiny travelers face the same dangers their parents did.

At home, they hide away. They will return to the coast in a few years!

baby crabs

Baby Christmas Island
Red Crab Return

mode of travel: walking

arriving
December/January:
rain forests

leaving
1 month after
hatching: coast

Glossary

breed—to produce babies

burrows—underground homes where Christmas Island red crabs live

crustaceans—animals that have several pairs of legs and hard outer shells

dry season—a time of year when it rains very little

hatch—to break out of an egg

larvae—baby crabs that have just hatched from eggs

mature—fully developed

migration—the act of moving from one place to another, often with the seasons

nimble—able to move quickly and with ease

predators—animals that hunt other animals for food

rain forest—a warm forest that has tall trees and receives a lot of rain

shields—pieces of armor that block someone or something from harm

wet season—a time of year when it rains a lot

To Learn More

AT THE LIBRARY
Davies, Monika. *How Far Home? Animal Migrations.*
Mankato, Minn.: Amicus Illustrated, 2019.

Katz Cooper, Sharon. *When Crabs Cross the Sand:
The Christmas Island Crab Migration.* North Mankato,
Minn.: Capstone, 2015.

Packham, Chris. *Amazing Animal Journeys.* New York,
N.Y.: Sterling Children's Books, 2016.

ON THE WEB
Learning more about
Christmas Island red crab
migration is as easy as 1, 2, 3.

1. Go to www.factsurfer.com.

2. Enter "Christmas Island red crab migration"
   into the search box.

3. Click the "Surf" button and you will see a
   list of related web sites.

With factsurfer.com, finding more information is
just a click away.

Index

The images in this book are reproduced through the courtesy of: KiltedArab, front cover (crab), pp. 5, 6; Pakhnyushchy, front cover (sky); Big Foot Productions, front cover (sand); Muzhik, front cover (gradient map); Stephen Belcher/ Minden Pictures/ SuperStock, pp. 4-5, 10; Nijethorpe, p. 7; John Tann/ Wikipedia, p. 8; Jean-Paul Ferrero/ Pantheon/ SuperStock, p. 9; Gary Tindale/ Flickr, pp. 10-11; Minden Pictures/ SuperStock, pp. 14, 16; Reinhard Dirscherl/ age fotostock/ SuperStock, pp. 17, 20, 20-21; Reinhard Dirscherl/ Mauritius/ SuperStock, p. 18 (top, bottom); WaterFrame/ Alamy, p. 19.